UNCOVER THE
MYSTERIES OF
ELECTRONICS:

A deep dive into
Voltage, Current, and Resistance!

John Och

Copyright © 2024 by John Och

All rights reserved.

No part of this publication may be reproduced, distributed, or transmitted in any form or by any means, including photocopying, recording, or other electronic or mechanical methods, without the prior written permission of the publisher, except in the case of brief quotations embodied in critical reviews and certain other noncommercial uses permitted by copyright law.

Table of Contents:

I. **Introduction** — 4
   - The Importance of Electronics Fundamentals
   - Overview of Voltage, Current, and Resistance

1. **Voltage Dynamics** — 5
   - Understanding Voltage
   - Voltage in Everyday Devices
   - Voltage in Power Systems
   - Practical Exercises

2. **Navigating Currents** — 17
   - Exploring Different Current Types
   - Current Flow in Electronic Circuits
   - Applications of Current Manipulation
   - Hands-On Activities

3. **Conquering Resistance Challenges** — 30
   - The Role of Resistance in Electronics
   - Resistance in Various Materials and Components
   - Optimizing Designs through Resistance
   - Case Studies and Problem-Solving Scenarios

4. **Real-World Applications** — 51
   - Voltage, Current, and Resistance in Action
   - Examples from Household Gadgets to Complex Systems
   - Bridging Theory with Practical Implementation
   - Project Ideas for Readers

5. **Empowering Problem-Solvers and Innovators** — 76
   - Developing a Problem-Solving Mindset
   - Innovations Fueled by Electronics Fundamentals
   - Building a Community of Enthusiasts
   - Encouraging Further Exploration

6. **Conclusion** — 89
   - Recap of Key Concepts
   - The Journey Towards Technological Excellence
   - Next Steps for Readers

Introduction

Electronics Fundamentals: Unlocking the Secrets of Voltage, Current, and Resistance"! Prepare to embark on a thrilling journey into the captivating world of electronics, where every flip of a switch unleashes a world of possibilities. From the sleek gadgets we cherish to the towering marvels of engineering, electronics are the heartbeat of our modern existence.

In this introduction, we beckon you to join us in uncovering the mysteries that lie beneath the surface of voltage, current, and resistance. Whether you're an intrepid explorer seeking knowledge or a seasoned enthusiast looking to deepen your understanding, this book promises to be your trusted companion.

As we delve into the core principles that govern electronic circuits, prepare to be captivated by practical insights, mesmerizing examples, and glimpses into the future of technology. By the journey's end, you'll emerge not only enlightened but empowered to wield the power of electronics with confidence and finesse. Let's embark on this electrifying odyssey together!

Chapter 1.

Voltage Dynamics: Unveiling the Power Behind Modern Technology

Understanding Voltage:

Voltage is a fundamental concept in the realm of electricity and electronics. Simply put, voltage represents the potential energy difference between two points in an electrical circuit. It's like the pressure in a water pipe, pushing electrons along a conductor.

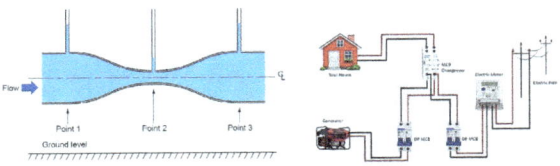

Fig.1.1 energy difference between two points

To understand voltage, it's essential to understand its measurement, which is typically in volts. This measurement indicates the amount of potential energy per unit charge. The higher the voltage, the greater the potential energy, and the more electrons will flow in a circuit.

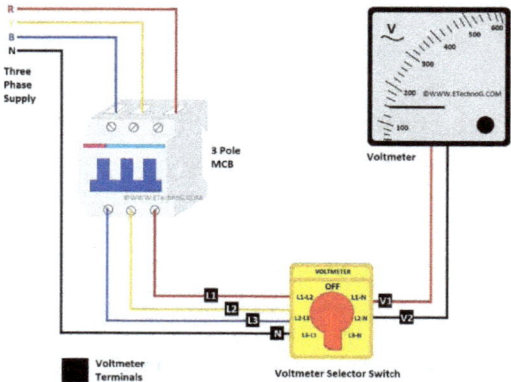

Fig1.2 voltage measurement

Voltage plays a crucial role in various aspects of daily life, from powering our electronic devices to driving industrial machinery. By understanding voltage, you gain insight into how electricity works and how to harness its power effectively.

Voltage is the driving force behind electrical circuits, providing the energy needed to power our modern world. By mastering the concept of voltage, you unlock a deeper understanding of electronics and pave the way for exploring more complex topics in electrical engineering.

Voltage in Everyday Devices:

Voltage powers a plethora of everyday devices that we rely on for convenience, communication, and entertainment. From the humble flashlight to the sophisticated smartphone, voltage plays a crucial role in enabling these devices to function.

In everyday devices, voltage serves as the energy source that powers electronic circuits. For example, in a smartphone, the battery provides a specific voltage level that drives the display, processor, and other components.

Fig1.3 smart device

Each component in the device operates at a specific voltage level, carefully regulated to ensure optimal performance and longevity.

Understanding voltage in everyday devices involves recognizing how different components utilize and respond to voltage. For instance, sensitive components like microprocessors require stable voltage levels to operate correctly, while others, such as LEDs, may require specific voltage levels for illumination.

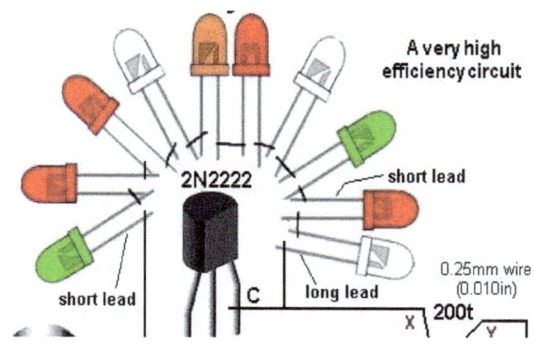

Fig1.4 LEDs

Furthermore, variations in voltage can affect device performance and lifespan. Overvoltage or undervoltage conditions can lead to component damage or malfunction, highlighting the importance of proper voltage regulation and management in device design and operation.

In essence, voltage in everyday devices is the lifeblood that powers our modern conveniences. By understanding how voltage is utilized and managed in these devices, we gain insight into their operation and maintenance, empowering us to make informed decisions about their use and care.

Voltage in Power Systems:

Voltage is essential to the production, transfer, and distribution of electrical energy in power systems. In order to generate, transmit, and distribute electricity from power plants to customers, power systems are enormous networks of interconnected components.

Fundamentally, voltage in power systems is the potential difference between two electrical circuit locations. Power is delivered to residences, companies, and industries via transmission lines and distribution networks, which are powered by this potential difference.

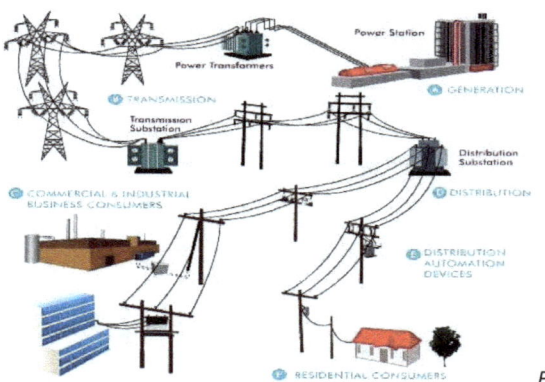

Fig 1.5 Power distribution system.

In power systems, voltage control is necessary to keep network voltage levels constant.

10

Variations in voltage can have a negative impact on electrical equipment performance and cause problems with power quality. To maintain consistent voltage conditions across the grid, electricity providers use a variety of voltage regulation tools, including mills and voltage regulators.

Power systems not only need to precisely control voltage, but they also need to precisely control voltage in order to satisfy the diverse demands of consumers. The grid cannot operate reliably and effectively unless it strictly adheres to voltage conditions while accounting for fluctuations in freight and generation.

Fig1.6 Power grid lines.

When everything is considered, voltage in power systems is one of the most important factors affecting the effectiveness, stability, and accountability of electrical distribution networks. Masterminds and drivers may build, run, and maintain energy grids to satisfy changing societal requirements while maintaining system integrity and safety by knowing the function that voltage plays in power systems.

Understanding Voltage Practical Exercises:

Practical exercises are invaluable tools for reinforcing theoretical knowledge and developing hands-on skills in understanding voltage. These exercises provide an opportunity for learners to apply concepts learned in the classroom or from textbooks in real-world scenarios. Here's an overview of the benefits and examples of voltage practical exercises:

Benefits:

Application of Theory: Practical exercises allow learners to apply theoretical concepts, such as Ohm's law and voltage calculations, in simulated or real-world situations.

Hands-On Experience: Hands-on experience with voltage measurement tools, such as multimeters and oscilloscopes, enhances understanding and proficiency.

Problem-Solving Skills: Practical exercises often involve troubleshooting circuits or analyzing voltage variations, helping learners develop critical thinking and problem-solving skills.

Skill Development: Performing voltage measurements, circuit simulations, and component testing enhances practical skills essential for careers in electrical engineering, electronics, and related fields.

Confidence Building: Successfully completing practical exercises builds confidence and reinforces understanding, motivating learners to tackle more complex challenges.

Examples of Voltage Practical Exercises

Voltage Measurement: Use a multimeter to measure the voltage across various components in a simple circuit, such as resistors and capacitors.

Fig1.7 multimeter

Voltage Divider Circuit: Build and analyze a voltage divider circuit to understand how different resistor values affect output voltage.

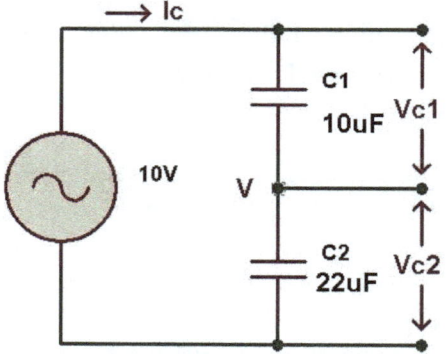

Fig1.8 voltage divider circuit

Circuit Simulation: Use circuit simulation software to design and simulate voltage divider circuits, exploring different configurations and component values.

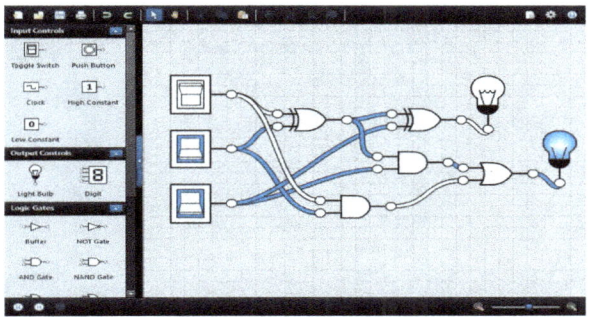

Fig1.9 circuit simulation

Troubleshooting: Identify and rectify voltage-related issues in malfunctioning circuits, such as open circuits, short circuits, or voltage drops.

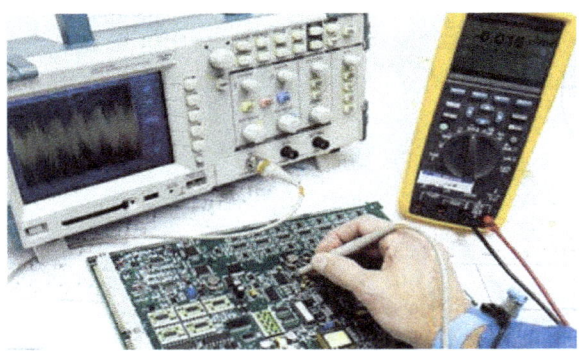

Fig1.10 Troubleshooting

Oscilloscope Analysis: Use an oscilloscope to analyze voltage waveforms in AC circuits, including amplitude, frequency, and phase relationships.

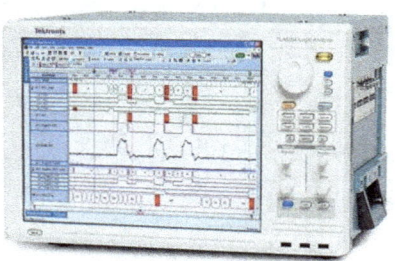

Fig1.11 Oscilloscope.

By engaging in practical exercises focused on voltage, learners gain hands-on experience, develop essential skills, and deepen their understanding of electrical concepts. These exercises not only complement theoretical learning but also prepare learners for real-world challenges in electrical engineering and electronics.

Chapter 2.

Navigating Current: Exploring the Path of Electrical Flow

Navigating current flow

Navigating current is akin to embarking on a journey through the dynamic landscape of electricity, where electrons dance through conductors, illuminating the path to understanding. Current, the flow of electric charge, is the driving force behind electrical systems, powering devices and enabling communication and functionality.

Understanding how to navigate current involves recognizing its various forms and pathways. Direct current (DC) flows steadily in one direction, like a calm river flowing downstream,

while alternating current (AC) oscillates back and forth, resembling the ebb and flow of ocean waves.

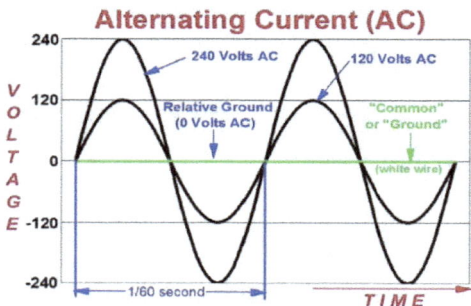

Fig2.1 Alternating current

Each type of current has its own unique characteristics and applications, shaping the design and operation of electrical systems.

Navigating current also entails tracing its path through electronic circuits, understanding how it flows from a source, such as a battery or power supply, through components, and back again. By analyzing circuit diagrams and understanding the principles of series and parallel circuits, we gain insight into how current behaves and interacts with different elements within a circuit.

Furthermore, navigating current involves exploring its applications in various fields, from powering household appliances to driving industrial machinery. By understanding how current can be manipulated and controlled, we can design and optimize electrical systems for efficiency, reliability, and safety.

In essence, navigating current is about understanding the flow of electrical energy and harnessing its power to create, innovate, and improve our world. By mastering the principles of current navigation, we can unlock new possibilities and drive progress in the field of electrical engineering and beyond.

Exploring Different Current Types:

Electricity comes in various forms, each with its own unique characteristics and applications. Exploring different current types allows us to understand how electrical energy behaves in different contexts.

Direct Current (DC): Direct current flows steadily in one direction, like a river flowing consistently downstream. It is commonly found in batteries and DC power supplies, providing a stable source of power for devices such as flashlights, calculators, and electronic gadgets.

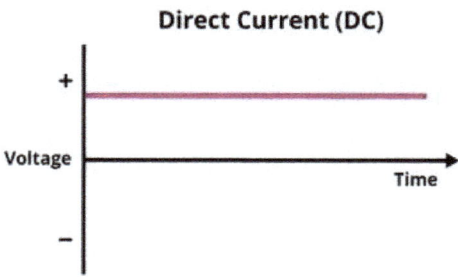

Fig2.2 Direct current (DC)

Alternating Current (AC): Alternating current oscillates back and forth in a periodic fashion, resembling the rhythmic rise and fall of ocean waves. AC is the type of current used in household electrical systems and power grids, delivering electricity to homes, businesses, and industries. Its ability to change direction allows for efficient transmission over long distances with minimal energy loss.

Understanding the differences between DC and AC is essential for designing and operating electrical systems effectively. While DC is suitable for low-power applications requiring a stable source of energy, AC is ideal for high-power transmission and distribution systems.

By exploring the diversity of current types, we gain insight into the principles of electrical flow and the role each type plays in powering our modern world. Whether it's lighting up our homes, charging our devices, or driving industrial machinery, the exploration of different current types illuminates the versatility and power of electricity.

Understanding Current Flow in Electronic Circuits

Current flow in electronic circuits is akin to tracing the journey of electric charge through a complex network of components and pathways. It is fundamental to comprehending the behavior and functionality of electronic devices and systems.

Source to Load: Current flows from a source of electrical energy, such as a battery or power supply, through conductive pathways within a circuit. It travels through various components, including resistors, capacitors, and transistors, before reaching its destination, known as the load.

Fig2.3 Battery power supply.

Closed Loop: In a closed circuit, current forms a continuous loop, flowing from the positive terminal of the source, through the circuit components, and back to the negative terminal of the source. This continuous flow of current is essential for powering devices and enabling functionality.

Fig2.4 closed circuit,

Series and Parallel Circuits: Current flow in electronic circuits can take two primary configurations: series and parallel. When parts of a circuit are joined end to end, a single path for current flow is created. In a parallel circuit, components are connected across multiple pathways, allowing current to divide and flow through each branch independently.

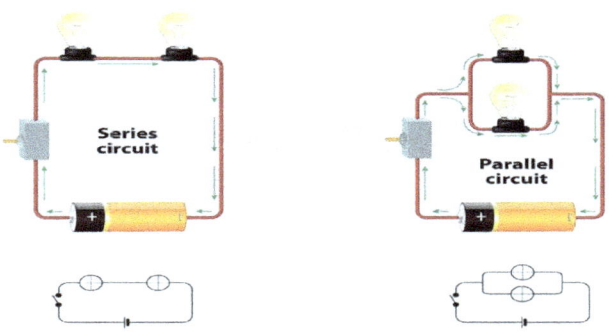

Fig2.5 Series and Parallel Circuits

Voltage and Resistance: Current flow in electronic circuits is governed by the principles of Ohm's Law, which relates voltage, current, and resistance. According to Ohm's Law ($V = IR$), the current (I) flowing through a circuit is directly proportional to the voltage (V) applied across it and inversely proportional to the resistance (R) encountered.

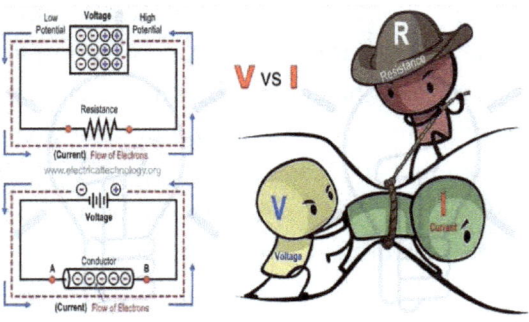

Fig2.6 Voltage and Resistance

Understanding current flow in electronic circuits is essential for designing, analyzing, and troubleshooting electrical systems. By tracing the path of electrical energy through circuits and understanding the principles that govern its behavior, we gain insight into how electronic devices operate and how they can be optimized for efficiency and performance.

Applications of Current Manipulation:
Current manipulation involves controlling the flow of electric charge in circuits to achieve specific outcomes. This versatile technique finds numerous applications across various industries and disciplines, driving innovation and progress in modern society.

Electronics: Current manipulation is fundamental to the operation of electronic devices, from smartphones and computers to televisions and radios. By controlling the flow of current through semiconductor components such as transistors and diodes, electronics engineers can create circuits that perform complex functions such as amplification, switching, and signal processing.

Fig2.7 Electronics

Power Generation and Distribution: In power generation and distribution systems, current manipulation is essential for managing electricity supply and demand efficiently.

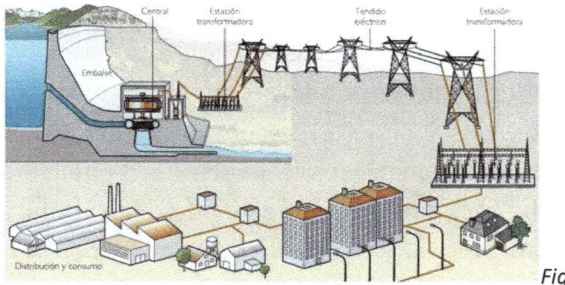

Fig 2.8 Power Generation and Distribution.

Power plants use generators to produce alternating current (AC), which is then manipulated through transformers, switches, and other control devices to match the requirements of consumers and ensure reliable and stable power delivery.

Electric Motors and Machinery: Electric motors and machinery rely on current manipulation to convert electrical energy into mechanical work. By controlling the flow of current through coils of wire in magnetic fields, electric

motors generate rotational motion, powering everything from industrial machinery and transportation systems to household appliances and electric vehicles.

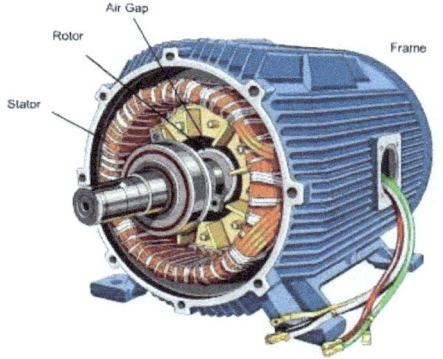

Partially Assembled Motor

Fig2.9 Electric Motors

Renewable Energy: In renewable energy systems such as solar panels and wind turbines, current manipulation is used to convert energy from natural sources into electrical power. Control systems regulate the flow of current to maximize energy capture and optimize system performance, enabling the widespread adoption of clean and sustainable energy technologies.

Fig2.10 Renewable Energy

Medical Devices: Current manipulation plays a crucial role in medical devices and treatments, such as pacemakers, defibrillators, and electrotherapy devices.

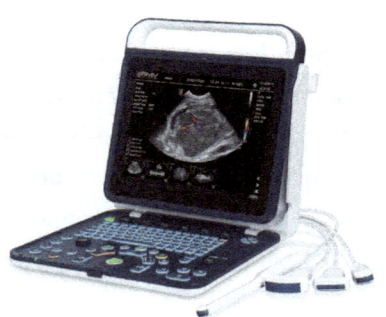

Fig2.11 ultrasound device

By precisely controlling the flow of current through tissues and organs, medical

28

professionals can diagnose and treat various medical conditions, providing patients with life-saving interventions and therapies.

Overall, current manipulation is a powerful tool that enables us to harness the potential of electricity for a wide range of applications. Whether it's powering our electronic devices, driving industrial machinery, or advancing medical treatments, current manipulation continues to drive innovation and improve our quality of life.

Chapter 3

Conquering the Resistance Challenge: Overcoming Obstacles in Electrical Circuits

In the world of electrical engineering, *resistance* is a common handicap that can stymie the inflow of electric current and hamper the performance of circuits. Conquering the resistance challenge involves understanding the nature of resistance and enforcing strategies to alleviate it's effects.

Understanding Resistance. Resistance is a property of material or components that opposes the inflow of electric current.

Fig3.1 Resistance.

It's measured in ohms(Ω) and is told by factors similar as the material's conductivity, length, andcross-sectional area. High resistance can affect in voltage drops, power losses, and heating of components.

Minimizing Resistance: One approach to conquering the resistance challenge is to minimize resistance within electrical circuits. This can be achieved by using conductive materials with low resistivity, similar as bobby or aluminum, and by reducing the length and adding thecross-sectional area of conductors.

Fig3.2 Semiconductors.

Ohm's Law and Resistance: Ohm's Law(V = IR) describes the relationship between voltage(V), current(I), and resistance(R) in a circuit. By applying Ohm's Law,

OMH'S LAW

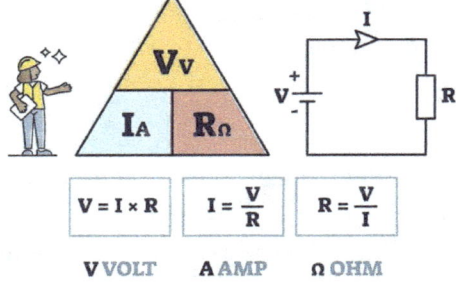

Fig3.3 Ohm's Law and Resistance

masterminds can calculate the voltage drop across resistive factors and determine the quantum of current flowing through them, allowing for the optimization of circuit design and performance.

Temperature effects: Resistance can vary with temperature, especially in conductive materials similar as semiconductors and thermistors.

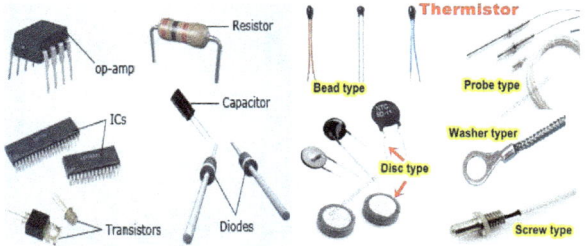

Fig3.4 Semiconductors.   Thermistors.

Conquering the resistance challenge involves considering temperature effects and enforcing thermal operation ways to insure stable and dependable operation of circuits.

Advanced techniques: Advanced techniques similar as feedback control, active factors(e.g., amplifiers), and signal exertion can be employed to overcome resistance challenges in complex electronic systems.

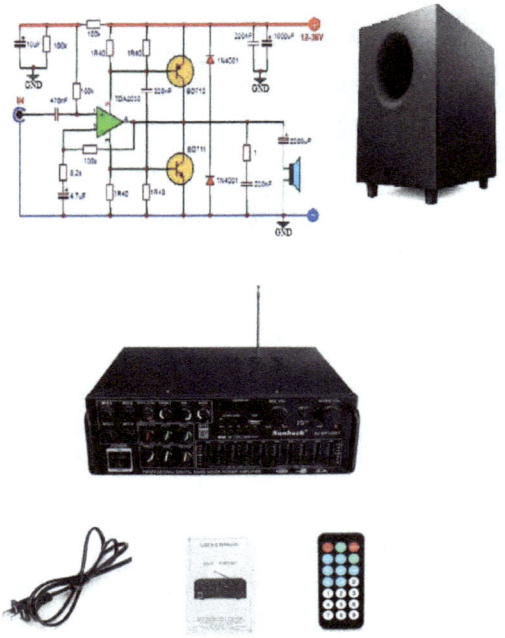

Fig3.5 amplifier

These techniques allow masterminds to stoutly acclimate voltage, current, and impedance to maintain asked performance situations and achieve specific design objectives.

By understanding the nature of resistance and enforcing effective mitigation strategies, masterminds can conquer the resistance challenge and design electrical circuits that are effective, dependable, and robust. This enables the development of innovative technologies and results that drive progress and ameliorate our quality of life.

The Role of Resistance in Electronics:

Resistance serves as a fundamental element in the world of electronics, playing a pivotal role in shaping the behavior and functionality of electrical circuits.

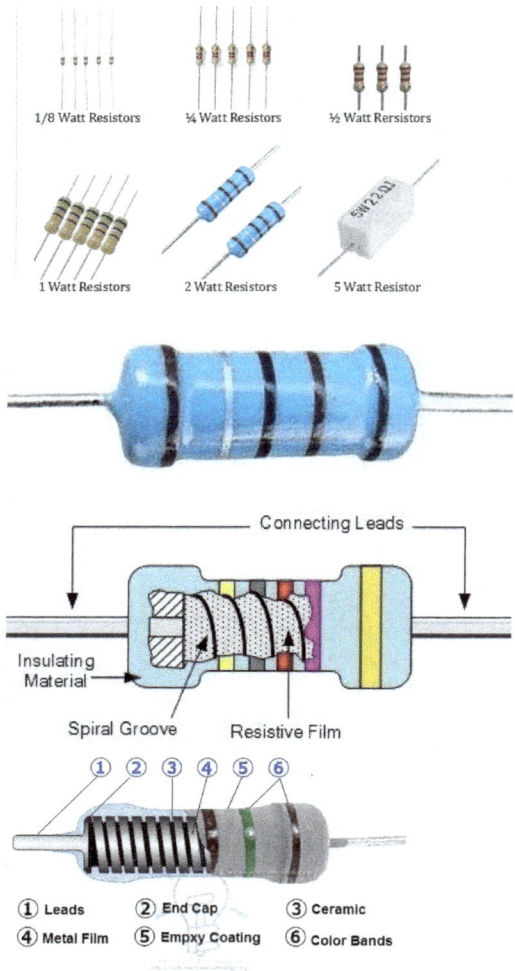

Fig3.6 resistors

Understanding its significance is essential for designing, analyzing, and optimizing electronic systems.

Limiting Current Flow: Resistance acts as a barrier to the flow of electric current within a circuit. It impedes the movement of electrons, converting electrical energy into heat in the process.

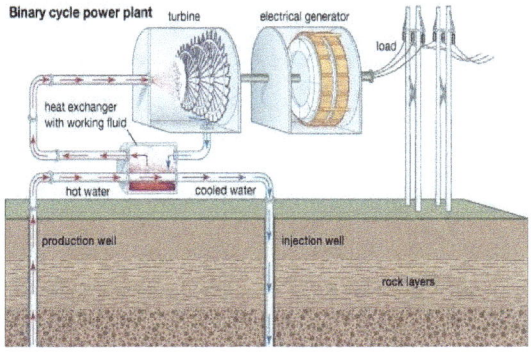

Fig3.7converting electrical energy into heat

This property allows resistance to control the amount of current flowing through various components, ensuring that circuits operate within safe and specified limits.

Voltage Division: In series circuits, resistance divides the applied voltage among multiple components, resulting in voltage drops across individual resistors. This phenomenon, known as voltage division, is utilized in voltage dividers and voltage measurement circuits to obtain specific voltage levels for various applications.

Current Regulation: Resistors are commonly used to regulate current flow in circuits. By placing a resistor in series with a load, engineers can limit the amount of current supplied to the load, protecting it from damage due to excessive current. This technique is widely employed in LED circuits, where resistors are used to control the brightness of LEDs and prevent overcurrent conditions.

Temperature Sensors: Certain materials exhibit a change in resistance with temperature, making them suitable for use as temperature sensors.

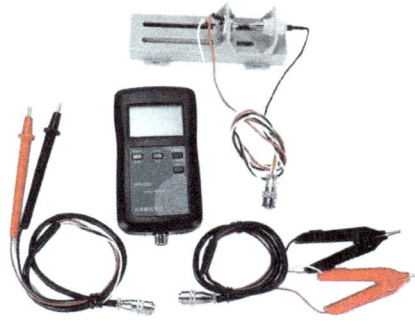

Fig3.8 temperature sensors.

Thermistors and resistance temperature detectors (RTDs) are examples of temperature-sensitive resistors commonly used in electronic circuits to measure and control temperature in applications such as thermostats, temperature controllers, and thermal protection systems.

Signal Processing: Resistors play a crucial role in signal processing circuits, where they are used to shape and manipulate electrical signals. In audio amplifiers, for example, resistors are employed in feedback networks to stabilize amplifier gain and improve overall performance.

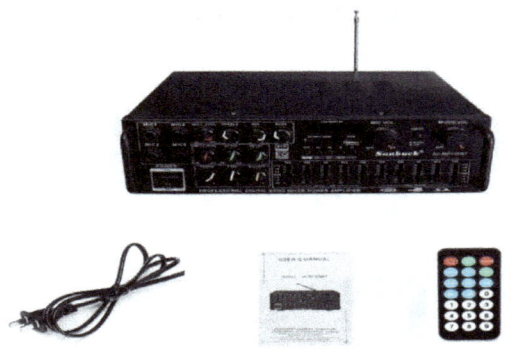

Fig3.9 Amplifier.

Additionally, resistors are used in filter circuits to attenuate or amplify specific frequency components of signals, allowing for selective signal processing.

Resistance is an indispensable component in electronics, contributing to the stability, safety, and functionality of electrical circuits. By understanding its role and properties, engineers can effectively design circuits to meet desired performance criteria and achieve specific design objectives.

Resistance in Various Materials and Components

Resistance varies depending on the material and element used in an electrical circuit. This is a brief overview of resistance in different materials and components.

Conductors: Materials with low resistance are known as conductors. Copper and aluminum are generally used conductors in electrical wiring due to their excellent conductivity.

Fig 3.9 Conductors.

These materials offer low resistance to the inflow of electric current, making them ideal for transmitting electricity with minimum powerloss.

Insulators: Materials similar as rubber, plastic, and glass, have veritably high resistance and are used to help the inflow of electric current.

Fig3.10 Insulators.

Insulators are pivotal for electrical safety, as they help to insulate conductive corridor and help electrical shocks.

Semiconductors: Semiconductors have resistance situations between those of conductors and insulators. Silicon and germanium are two common semiconductor materials used in electronic factors similar as diodes and transistors.

Fig 3. 11 Semiconductors.

The resistance of semiconductors can be controlled and manipulated through doping and other ways, allowing for the creation of electronic bias with specific electricalcharacteristics.

Resistors: Resistors are components specifically designed to give resistance in an electrical circuit.

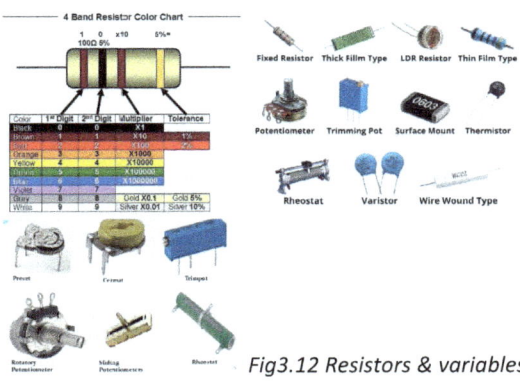

Fig3.12 Resistors & variables

They're generally made from materials as carbon composition, essence film, or line crack around a ceramic core. Resistors come in colorful resistance values and power conditions, allowing engineers to precisely control current

43

inflow and voltage sensitive situations in circuits.

Thermistors: Thermistors are temperature-resistors whose resistance varies with temperature.

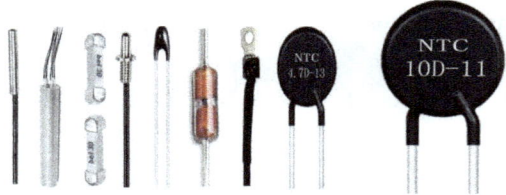

Fig3.13 Thermistors

They're generally used in temperature seeing and compensation operations, similar as thermostats, temperature regulators, and thermal protection systems.

Light-dependent Resistors(LDRs): LDRs are resistors whose resistance changes in response to changes in light intensity.

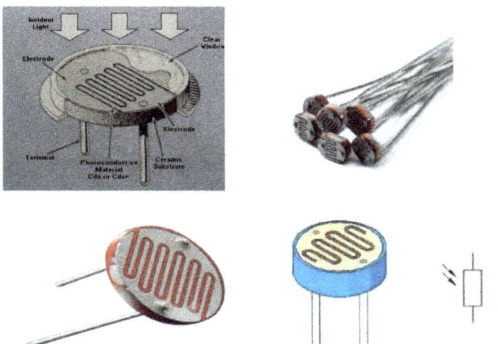

Fig1.

14 Light-dependent Resistors.

They're used in light sensing applications, similar as automatic lighting control systems and photographic lightmeters.

Understanding the resistance parcels of different materials and factors is essential for designing and assaying electrical circuits. By opting the applicable materials and factors, engineers can insure optimal circuit performance and trustability in colorful operations.

Optimizing Designs through Resistance

Optimizing designs through resistance involves strategically managing resistance positions within electrical circuits to achieve desired performance characteristics. This is how resistance optimization plays a pivotal part in circuit design.

Power Efficiency: By minimizing resistance in conductive paths, engineers can reduce power losses and better the overall effectiveness of electrical systems. This is particularly important in high- power operations, similar as power distribution networks and electric vehicles, where minimizing resistance leads to increased energy savings and enhanced system performance.

Signal Integrity: In signal processing circuits, optimizing resistance helps maintain signal integrity and dedication. By precisely choosing resistor values and minimizing parasitic resistance, engineers can help signal declination and insure accurate transmission of data andinformation.

Thermal Management: Resistance optimization is essential for managing heat dispersion within electronic bias. By controlling the inflow of current and minimizing resistance in critical factors, engineers can prevent overheating and thermal runway, ensuring the reliability and life of electronic systems.

Noise Reduction: High resistance can contribute to signal noise and hindrance in electronic circuits. By optimizing resistance positions and minimizing unwanted resistance sources, similar as inadequately designed connections or rightly shielded factors, engineers can reduce noise situations and better the signal- to- noise rate in sensitive applications.

Component Selection: Choosing the right resistive factors, similar as resistors and thermistors, is pivotal for optimizing circuit performance. By choosing factors with applicable resistance values, power conditions, and temperature portions, engineers can adjust circuit designs to meet specific conditions and achieve desired performance characteristics.

Cost Optimization: Resistance optimization also involves balancing performance conditions with cost considerations. By precisely choosing materials and factors and optimizing circuit layouts, engineers can achieve the asked position of performance while minimizing manufacturing costs and maximizing overall value.

Optimizing designs through resistance is a multifaceted process that involves balancing performance, effectiveness, trustability, and cost considerations. By strategically managing resistance situations within electrical circuits, engineers can achieve optimal results and deliver innovative results that meet the requirements of ultramodern technology and assiduity.

Case Studies and Problem-Solving Scenarios.

Case studies and problem-solving scenarios provide valuable insights into real-world applications of resistance optimization and circuit design. Here are a few examples

Power Distribution Network Optimization: In a large-scale power distribution network, engineers encounter challenges related to power losses and efficiency. By analyzing the network topology and identifying areas of high resistance, engineers can optimize conductor sizes, routing paths, and connection methods to minimize resistance and improve overall system efficiency.

Signal Integrity Enhancement: In a high-speed data transmission system, signal integrity is critical for reliable communication. Engineers face challenges related to signal degradation and noise interference caused by resistance in signal paths and connectors. By implementing impedance matching techniques, signal conditioning circuits, and low-resistance interconnects, engineers can optimize signal integrity and ensure accurate data transmission.

Thermal Management in Electronic Devices: In a compact electronic device, such as a smartphone or laptop, thermal management is essential for preventing overheating and component failure. Engineers encounter challenges related to high-power components generating excess heat and increasing resistance in conductive paths. By optimizing component placement, heat sinks, and airflow patterns, engineers can minimize resistance and improve thermal dissipation, ensuring reliable device operation.

Noise Reduction in Audio Systems: In an audio amplification system, engineers face challenges related to noise interference and distortion caused by resistance in signal paths and components. By using low-noise amplifiers, shielded cables, and noise filtering techniques, engineers can optimize circuit design to minimize resistance-induced noise and achieve high-fidelity audio reproduction.

Cost-Effective Circuit Design: In a consumer electronics product, such as a smart home device or wearable technology, engineers must balance performance requirements with cost considerations. By optimizing circuit design for efficiency, reliability, and manufacturability, engineers can achieve the desired level of performance while minimizing manufacturing costs and maximizing overall value for customers.

Case studies and problem-solving scenarios provide engineers with valuable opportunities to apply theoretical knowledge to real-world challenges, fostering innovation, creativity, and continuous improvement in circuit design and optimization.

Chapter 4:

Advanced Applications of Electrical Concepts

Voltage, current, and resistance are fundamental concepts in electricity and electronics, and they find numerous real-world applications in various fields. Here are some examples of how these concepts are applied in everyday life

Electrical Power Grids: In electrical power grids, voltage is used to transmit electrical energy over long distances from power plants to homes and businesses.

Fig4.1 High voltage grid lines.

High-voltage transmission lines minimize power losses during transmission, while transformers step down voltage levels to safer and more usable levels for distribution to end-users. Current flows through power lines to deliver electricity to consumers, and resistance in transmission lines affects power losses and efficiency.

Household Appliances: Household appliances such as refrigerators, washing machines, and televisions operate using voltage, current, and resistance.

Fig4.2 Household appliances

Voltage from electrical outlets powers these devices, while current flows through circuits to provide the necessary energy for operation.

52

Resistance in components such as heating elements regulates the flow of current and controls the device's function.

Electronics: In electronic devices such as stphones, laptops, and digital cameras, voltage, current, and resistance play crucial roles. mar

Fig4.3 Electronics

Voltage from batteries or power sources powers electronic circuits, while current flows through components such as resistors, capacitors, and transistors to perform various functions.

Fig4.4 Batteries

Resistance in circuit elements determines the behavior of electronic devices and helps control parameters such as signal amplitude, frequency, and power consumption.

Automotive Systems: In vehicles, voltage, current, and resistance are essential for various systems and components.

Fig4.4 Automotive Systems

The vehicle's electrical system operates at a specific voltage level provided by the battery and alternator, powering systems such as lights, ignition, and entertainment. Current flows through wires and components to operate various electrical systems, while resistance in components such as spark plugs and sensors affects engine performance and fuel efficiency.

Medical Devices: Medical devices such as pacemakers, defibrillators, and electrocardiographs rely on voltage, current, and resistance for diagnosis and treatment.

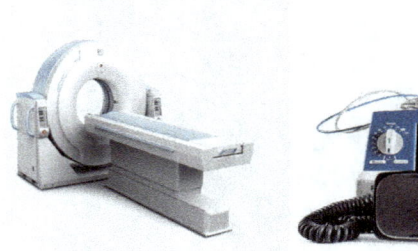

Fig4.5 Medical Devices

Voltage from batteries powers these devices, while current flows through electrodes and sensors to monitor physiological signals and deliver therapeutic interventions. Resistance in conductive pathways affects signal quality and device performance, ensuring accurate diagnosis and treatment of medical conditions.

These examples illustrate how voltage, current, and resistance are integral to numerous applications in our daily lives, from powering household appliances to enabling advanced technologies in healthcare and transportation. Understanding these concepts is essential for engineers, technicians, and consumers alike to effectively design, operate, and troubleshoot electrical and electronic systems.

Applications of Voltage, Current, and Resistance in Nanoelectronics

Applications of voltage, current, and resistance extend beyond basic electrical principles to encompass cutting-edge technologies and innovative solutions. Here are some advanced applications

Power Electronics: In high-power applications such as electric vehicles, renewable energy systems, and industrial automation, advanced power electronics utilize voltage converters, inverters, and rectifiers to efficiently manage voltage and current levels.

Fig4.6 Voltage regulator.   Inverter.

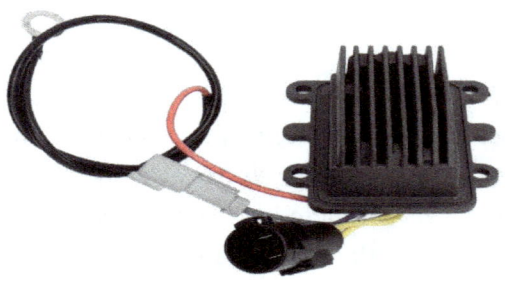

Fig4.7 Voltage rectifier.

These systems optimize energy conversion, minimize losses, and enhance overall system performance.

Nanoelectronics: Nanoscale devices are tiny machines or systems that operate on a incredibly small scale, typically measured in nanometers (nm). A nanometer is one billionth of a meter in size.. These devices are designed to perform specific tasks, such as: sensing and detecting, energy harvesting and storage, electronics and computing, medical applications (e.g., drug delivery, imaging), robotics and automation. Examples of nanoscale devices include: nanorobots, nanosensors, nanoantennas, anogenerators, quantum dots (tiny particles used for imaging and sensing)

The development of nanoscale devices requires advanced materials and techniques, such as: nanofabrication, molecular self-assembly, scanning probe microscopy, nanolithography

Nanoscale devices have the potential to revolutionize various fields, including healthcare, energy, and electronics, due to their unique properties and abilities..

Superconducting Technology: A superconductor is a material that can conduct electricity with zero resistance, meaning it can carry electrical current with perfect efficiency and without losing any energy.

Fig4.8 Superconducting Technology

This unique property allows superconductors to conduct electricity with no resistance, expel magnetic fields, maintain electrical current indefinitely. Superconductors have many potential applications, including: power transmission and storage, medical imaging and treatment (MRI, NMR), high-speed transportation (maglev trains), energy-efficient devices (motors, generators), advanced electronics and sensors.

Some common superconducting materials include: niobium (Nb), aluminum (Al), tin (Sn), lead (Pb), yttrium barium copper oxide (YBCO). Superconductors are typically very cold, requiring temperatures near absolute zero (-273.15°C) to exhibit their unique properties. However, researchers are working to develop materials that can superconduct at higher temperatures, making them more practical for everyday use.

Quantum Computing: In quantum computing, voltage, current, and resistance govern the behavior of quantum bits (qubits) and control signals in quantum circuits. Advanced quantum hardware architectures, such as superconducting qubits and trapped ions, exploit quantum phenomena to perform complex calculations with exponentially greater computational power than classical computers.

Photonics and Optoelectronics: In photonics and optoelectronics, voltage and current control the generation, modulation, and detection of light signals in advanced devices such as lasers, photodetectors, and optical fibers.

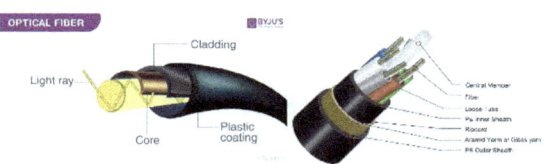

Fig4.9 Fiberoptic cable.

These technologies enable high-speed data transmission, optical sensing, and imaging applications in telecommunications, biomedical imaging, and quantum communication.

Neuromorphic Computing: Inspired by the structure and function of the human brain, neuromorphic computing systems utilize voltage-controlled synaptic connections and spiking neurons to perform cognitive tasks with high efficiency and parallelism. These advanced systems offer promising avenues for artificial intelligence, pattern recognition, and brain-inspired computing applications.

Flexible and Wearable Electronics: In flexible and wearable electronics, voltage, current, and resistance are manipulated in thin-film transistors, organic semiconductors, and stretchable conductive materials to create lightweight, bendable, and conformable devices. These advanced electronics enable applications such as flexible displays, electronic skins, and health monitoring systems.

Energy Harvesting and Storage: Voltage, current, and resistance are key parameters in energy harvesting and storage technologies such as thermoelectric generators, piezoelectric materials, and supercapacitors. Advanced energy harvesting systems convert waste heat, mechanical vibrations, and ambient light into electrical energy, while energy storage systems store and release energy with high efficiency and cycle life.

These advanced applications demonstrate the versatility and importance of voltage, current, and resistance in enabling transformative technologies across various domains, from nanoelectronics and quantum computing to photonics and energy systems.

Examples from Household Gadgets to Complex Systems

From smartphones to satellite communication, electronics fundamentals power everyday appliances and complex systems also. Whether optimizing battery life in smartphones or controlling discharges in electric vehicles, voltage, current, and resistance are the foundation of invention, effectiveness, and safety across a wide range of operations.

Household Gadgets:

Smartphones: Understanding voltage, current, and resistance is crucial for optimizing battery life and ensuring the safe operation of smartphone components.

Fig4.10 Smartphones

LED Light Bulbs: Voltage and current regulation are essential for controlling the brightness of

LED bulbs and extending their lifespan.

Fig4.11 LED Light Bulbs

Microwaves: Voltage regulation and current control are critical for the safe and efficient operation of microwave ovens.

Fig4.12 Microwaves

Automotive Electronics:

Electric Vehicles (EVs): Electronics fundamentals play a key role in managing the voltage and current flow in EV batteries, motor controllers, and charging systems.

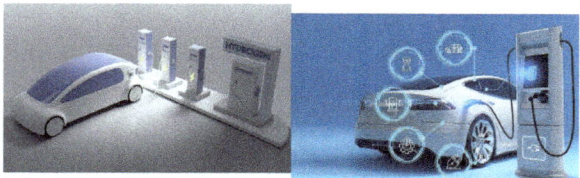

Fig4.13 Electric Vehicles

Engine Control Units (ECUs): Voltage, current, and resistance measurements are vital for monitoring and controlling various engine parameters, optimizing fuel efficiency, and reducing emissions.

Renewable Energy Systems:
Solar Photovoltaic (PV) Systems: Understanding voltage, current, and resistance is essential for designing and optimizing solar panel arrays, inverters, and battery storage systems.

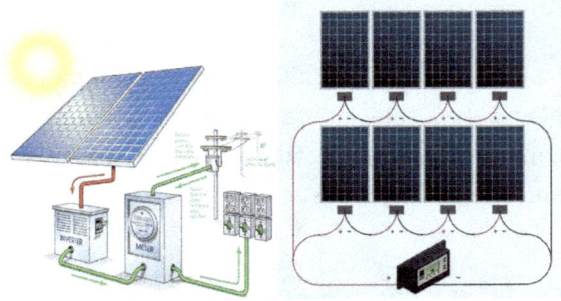

Fig1.14 Solar Photovoltaic

Wind Turbines: Electronics fundamentals are crucial for controlling the voltage and current output of wind turbines, ensuring efficient power generation and grid integration.

Fig4.15 Wind Turbines.

Communication Systems:

Wireless Networks: Voltage regulation and current management are critical for maintaining stable power supplies to wireless base stations and network infrastructure.

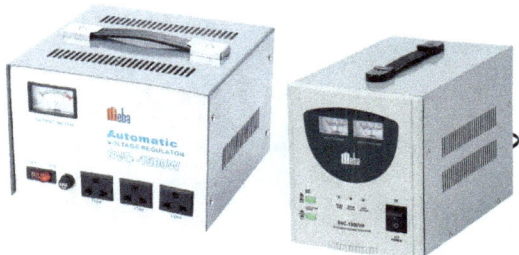

Fig4.16 stable power supplies

Satellite Communication Systems: Electronics fundamentals are essential for designing and optimizing satellite communication payloads, ensuring reliable transmission and reception of signals.

Fig4.17 Satellite Communication

Medical Devices:

Implantable Medical Devices: Voltage and current management are vital for powering and controlling implantable medical devices such as pacemakers, defibrillators, and insulin pumps.

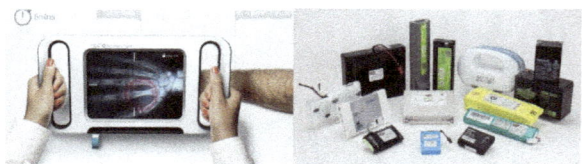

Fig4.18 Implantable Medical Devices

Diagnostic Equipment: Electronics fundamentals play a crucial role in the design and operation of medical diagnostic equipment

68

such as MRI machines, X-ray systems, and ultrasound devices.

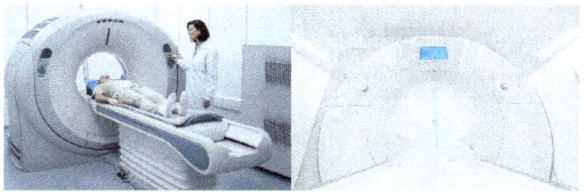

Fig4.19 Diagnostic Equipment.

By understanding and applying electronics fundamentals, engineers and technicians can develop innovative solutions, optimize performance, and ensure the reliability and safety of a wide range of electronic devices and systems across various industries.

Bridging Theory with Practical Implementation

Bridging theory with practical implementation is essential for applying electronics fundamentals effectively in real-world scenarios. Here are some examples:

Circuit Design: Understanding theoretical concepts like Ohm's Law and Kirchhoff's Laws enables engineers to design circuits that meet specific performance criteria.

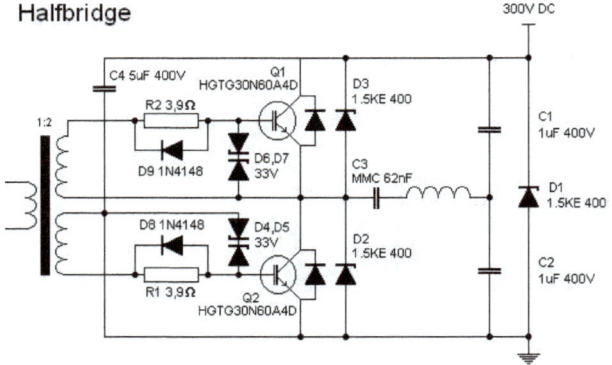

Fig4.20 *Circuit Design*

Practical implementation involves selecting appropriate components, layout considerations, and testing prototypes to ensure functionality.

Power Electronics: Theoretical knowledge of voltage, current, and resistance informs the design of power electronic converters, such as DC-DC converters and inverters. Practical implementation involves component selection, thermal management, and electromagnetic interference (EMI) mitigation to achieve desired performance and reliability.

Embedded Systems: Theoretical understanding of digital electronics principles, such as logic gates and flip-flops, enables engineers to design embedded systems for various applications.

Fig4.21 Embedded Systems:

Practical implementation involves programming microcontrollers, interfacing with sensors and actuators, and testing system functionality in real-world environments.

Signal Processing: Theoretical knowledge of analog and digital signal processing principles underpins the design of communication systems, audio equipment, and medical devices. Practical implementation involves designing filters, implementing algorithms in software or hardware, and optimizing system performance for specific applications.

Control Systems: Theoretical understanding of control theory principles, such as feedback loops and PID controllers, informs the design of control systems for processes ranging from temperature regulation to robotic motion control.

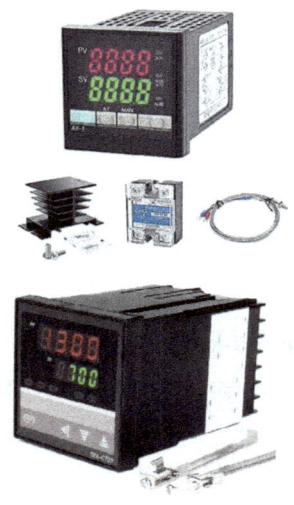

Fig4.22

Control Systems

Practical implementation involves tuning controller parameters, implementing control algorithms in hardware or software, and testing system response under various operating conditions.

By bridging theory with practical implementation, engineers can develop innovative solutions, optimize system performance, and troubleshoot issues effectively in real-world applications.

Real-world project ideas that readers can explore

Real-world project ideas that readers can explore to apply their knowledge of current, voltage, and resistance include:

Smart Home Energy Monitoring System:

Design and build a system to monitor energy usage in a smart home using current sensors to measure electrical currents flowing through different appliances.

Use voltage sensors to measure the voltage levels across various circuits and devices.

Implement data logging and visualization techniques to display energy consumption trends and identify opportunities for energy savings.

Solar-Powered Charger:

Develop a solar-powered charger for mobile devices using photovoltaic panels to convert sunlight into electrical energy.

Use voltage regulators to maintain a stable output voltage suitable for charging smartphones and other portable electronics.

Incorporate battery management circuitry to store excess energy for use during periods of low sunlight.

Variable Power Supply:

Construct a variable power supply capable of providing adjustable output voltages and currents for testing and prototyping electronic circuits.

Include voltage and current monitoring features to ensure accurate control and protection against overloads.

Implement feedback control mechanisms to regulate the output voltage and current based on user-defined settings.

Resistance-Based Sensor Network:

Develop a sensor network using resistive sensors to measure physical parameters such as temperature, humidity, and pressure in a home or industrial environment.

Interface the sensors with microcontrollers or single-board computers to digitize and transmit sensor readings wirelessly.

Build a centralized monitoring and control system to analyze sensor data, detect anomalies, and trigger alerts or automated responses.

Electric Vehicle Charging Station:
Design and construct an electric vehicle (EV) charging station capable of delivering high-power charging currents to recharge EV batteries quickly.

Incorporate voltage and current monitoring capabilities to ensure safe and efficient charging operations.

Implement smart charging features to optimize charging schedules based on grid conditions and user preferences.

These project ideas provide readers with opportunities to apply their understanding of current, voltage, and resistance in practical contexts, fostering creativity, problem-solving skills, and hands-on experience in electronics.

Chapter 5

Empowering problem-solvers and innovators

Empowering problem-solvers and innovators is at the heart of mastering electronics fundamentals. Here's how:

Critical Thinking Skills: Understanding concepts like voltage, current, and resistance fosters analytical thinking, enabling individuals to dissect complex problems and devise creative solutions.

Practical Problem-Solving: Armed with a deep understanding of electronics fundamentals, individuals can tackle real-world challenges with confidence, whether it's troubleshooting a malfunctioning device or optimizing the efficiency of an electrical system.

Innovation: Electronics fundamentals provide the building blocks for innovation, empowering individuals to develop groundbreaking technologies that push the boundaries of what's possible. From renewable energy solutions to advanced medical devices, the possibilities are endless.

Collaboration: By mastering electronics fundamentals, individuals become valuable collaborators, able to contribute their expertise to interdisciplinary teams working on projects spanning fields like robotics, telecommunications, and renewable energy.

Continuous Learning: Electronics is a dynamic field, constantly evolving with new technologies and advancements. By mastering fundamentals, individuals lay the foundation for lifelong learning, staying abreast of the latest developments and driving innovation forward.

Ultimately, mastering electronics fundamentals empowers individuals to become not just consumers of technology, but creators and innovators, shaping the future of our increasingly interconnected world.

Developing a problem-solving mindset

Developing a problem-solving mindset is crucial for success in any field, including electronics.

Guides to developing a problem- working mindset

Embrace Challenges. Rather of stewing challenges, see them as openings to learn and grow. Challenges help us develop adaptability and expand our chops. Approach each problem with a positive station and the belief that you can overcome it.

Break Down Problems. Complex problems can feel inviting at first regard. Divide them up into smaller, easier-to-manage tasks. This makes it easier to attack each aspect of the problem methodically and prevents feeling overwhelmed.

Ask Questions. Cultivate curiosity by asking questions about the problem at hand. Seek to understand the root causes and underpinning principles behind the problem. Asking questions helps you gain clarity and identify implicit results.

Suppose Creatively. Explore different perspectives and consider unconventional approaches to problem- working. Do not be hysterical to suppose outside the box and try new ideas. Creativity frequently leads to innovative results that others may not have considered.

Persist in the Face of Failure. Failure is a natural part of the problem- working process. Rather of viewing failure as a reversal, see it as an occasion to learn and grow. Dissect what went wrong, excerpt assignments from the experience, and use that knowledge to ameliorate your approach.

Seek Feedback. Do not vacillate to seek feedback from others, whether it's peers, instructors, or experts in the field. Formative feedback provides precious perceptivity and helps you see eyeless spots in your thinking. Be open to review and use it as a tool for enhancement.

Reiterate and Ameliorate. Continuously upgrade your results grounded on feedback and new information. Borrow a mindset of nonstop enhancement, seeking to make incremental progress over time. Do not settle for the first result that comes to mind; rather, reiterate and upgrade until you find the stylish possible outcome.

By embracing challenges, breaking down problems, asking questions, allowing creatively, persisting in the face of failure, seeking feedback, and repeating on results, you can develop a strong problem- working mindset. With this mindset, you will be better equipped to overcome obstacles.

Innovations Fueled by Electronics Fundamentals

Electronics fundamentals, encompassing concepts like voltage, current, and resistance, serve as the cornerstone for countless innovations that have revolutionized industries and transformed our daily lives. From the development of semiconductor technology to the advent of the Internet of Things (IoT), electronics fundamentals continue to fuel a wave of innovation across diverse sectors.

Semiconductor Revolution: Electronics fundamentals laid the groundwork for the semiconductor revolution, which paved the way for the miniaturization of electronic components and the birth of the modern computing era.

Innovations such as transistors, integrated circuits (ICs), and microprocessors, built upon principles of voltage, current, and resistance, have enabled the creation of increasingly powerful and compact electronic devices.

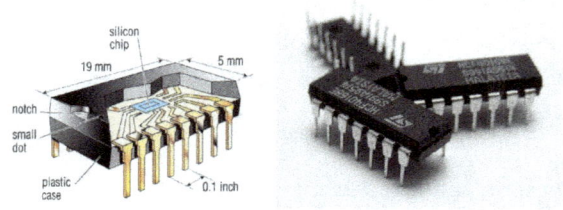

Fig 5.1 integrated circuits (ICs)

Digital Revolution: The digital revolution, driven by advancements in electronics fundamentals, has transformed communication, entertainment, and commerce on a global scale.

Innovations like digital signal processing (DSP), data storage technologies, and high-speed communication networks have reshaped how we access, share, and process information in the digital age.

Renewable Energy Technologies: Electronics fundamentals play a critical role in the development and deployment of renewable energy technologies such as solar photovoltaics (PV) and wind turbines.

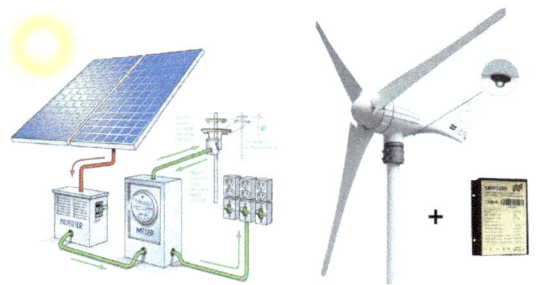

Fig5.2 Renewable Energy Technologies

Innovations in power electronics, energy storage systems, and grid integration solutions leverage voltage, current, and resistance principles to harness renewable energy sources and reduce reliance on fossil fuels.

Healthcare Innovations: In the field of healthcare, electronics fundamentals enable innovations such as medical imaging devices, implantable medical devices, and wearable health monitors.

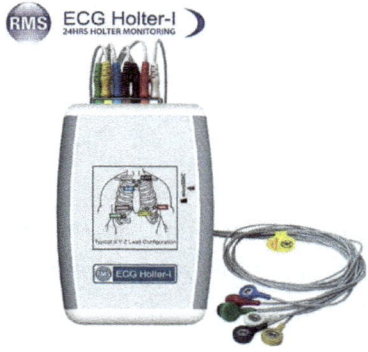

Fig5.3 ECG machine

Technologies like electrocardiography (ECG), magnetic resonance imaging (MRI), and pacemakers rely on precise control of voltage, current, and resistance to diagnose and treat medical conditions.

Internet of Things (IoT): The Internet of Things (IoT) ecosystem, comprising interconnected devices and sensors, is fueled by electronics fundamentals.

Innovations in sensor technology, wireless communication protocols, and embedded systems leverage voltage, current, and resistance principles to enable seamless connectivity and data exchange in IoT applications.

Robotics and Automation: Robotics and robotization technologies, driven by advancements in electronics fundamentals, are revolutionizing manufacturing, husbandry, and transportation.

Fig5.4 Robot and drone

Innovations similar as artificial robots, independent vehicles, and drones calculate on precise control of electrical signals to perform tasks efficiently andautonomously.

Inventions fueled by electronics fundamentals have reshaped our world, driving progress in areas ranging from calculating and communication to energy and healthcare. As technology continues to evolve, the principles of voltage, current, and resistance will remain at the heart of unborn inventions, empowering us to address global challenges and shape a further connected and sustainable future.

Building a Community of Enthusiasts

Building a community of enthusiasts is essential for fostering collaboration, sharing knowledge, and inspiring passion for electronics.

Online Forums and Social Media: Establish online forums and social media groups dedicated to electronics enthusiasts. These platforms provide a space for individuals to connect, share ideas, ask questions, and seek advice from peers.

Workshops and Meetups: Organize workshops, meetups, and seminars where enthusiasts can come together to learn new skills, collaborate on projects, and network with like-minded individuals. These events foster a sense of camaraderie and create opportunities for hands-on learning.

Collaborative Projects: Encourage collaborative projects that bring enthusiasts together to work towards a common goal. Whether it's building a robot, designing a circuit, or participating in a hackathon, collaborative projects foster teamwork and creativity.

Mentorship Programs: Establish mentorship programs where experienced enthusiasts can mentor and guide newcomers in their journey. Mentorship programs provide valuable support and encouragement, helping individuals overcome challenges and achieve their goals.

Recognition and Rewards: Recognize the contributions of community members through awards, certificates, and shout-outs. Celebrating achievements and milestones fosters a sense of pride and motivation within the community.

Encouraging Further Exploration

Encouraging further exploration is crucial for inspiring continuous learning and growth among enthusiasts. Here's how to encourage individuals to explore new areas and push the boundaries of their knowledge:

Continuing Education: Provide resources and opportunities for continuing education, such as online courses, workshops, and seminars. Encourage enthusiasts to expand their knowledge and skills in areas beyond their comfort zone.

Project Challenges: Organize project challenges that encourage enthusiasts to tackle new problems and explore innovative solutions. Whether it's a design competition or a coding challenge, project challenges provide a platform for enthusiasts to test their skills and creativity.

Guest Speakers and Experts: Invite guest speakers and specialists to impart their knowledge and expertise to the public. Guest lectures, webinars, and panel discussions provide valuable perspectives and inspire individuals to explore new ideas and technologies.

Hands-On Learning: Emphasize hands-on learning experiences that allow enthusiasts to apply theoretical knowledge in practical contexts. Whether it's building a prototype, conducting experiments, or troubleshooting a circuit, hands-on learning fosters a deeper understanding and appreciation for electronics.

Networking Opportunities: Facilitate networking opportunities with industry professionals, researchers, and other communities. Networking events, job fairs, and industry conferences provide opportunities for enthusiasts to learn about new trends, connect with potential mentors, and explore career pathways in electronics.

By implementing these strategies, we can build a vibrant and inclusive community of enthusiasts who are passionate about electronics and committed to lifelong learning, collaboration, and innovation. Together, we can inspire individuals to explore new frontiers, push the boundaries of their knowledge, and make meaningful contributions to the world of electronics and beyond.

Chapter 6

Recap of Key Concepts

Throughout this ebook, we've explored the foundational principles of electronics, including voltage, current, and resistance. These concepts serve as the building blocks for understanding how electronic devices and systems function. We've learned how voltage represents the electrical potential difference between two points, how current is the flow of electric charge through a conductor, and how resistance opposes the flow of current. By mastering these key concepts, readers have gained a deeper understanding of electronics and are better equipped to tackle real-world challenges and innovate in their respective fields.

The Journey Towards Technological Excellence:

The journey towards technological excellence is a continuous pursuit fueled by curiosity, passion, and dedication. It's about pushing the boundaries of what's possible, challenging conventional thinking, and driving innovation forward. As we embark on this journey, we're guided by a commitment to excellence, a thirst for knowledge, and a desire to make a positive impact on the world. Along the way, we'll encounter obstacles, setbacks, and moments of uncertainty, but it's these challenges that ultimately propel us forward, fueling our determination to succeed. With each new discovery, breakthrough, and achievement, we inch closer to realizing our full potential and shaping a future defined by technological advancement and human ingenuity.

Next Steps for Readers

As readers continue their exploration of electronics fundamentals, there are several key steps they can take to further their knowledge and expertise.

Continued Learning: Embrace a mindset of lifelong learning and seek out opportunities to deepen your understanding of electronics concepts. Whether through online courses, textbooks, or hands-on projects, there's always more to learn and discover.

Practical Application: Apply your newfound knowledge to real-world projects and challenges. Whether it's building your own electronic circuits, designing innovative solutions, or troubleshooting technical issues, hands-on experience is invaluable for solidifying your understanding and honing your skills.

Networking and Collaboration: Connect with other aficionados, professionals, and electronics specialists. Join online communities, attend networking events, and seek out mentorship opportunities to learn from others and expand your professional network.

Exploration of Specialized Areas: Explore specialized areas within electronics that align with your interests and career goals. Whether it's renewable energy, robotics, embedded systems, or telecommunications, there are countless avenues to explore and opportunities to make a meaningful impact.

Professional Development: Invest in your professional development by pursuing certifications, attending conferences, and participating in continuing education programs. Stay up-to-date with the latest advancements and trends in the field to remain competitive and position yourself for success.

By taking these next steps, readers can continue their journey towards technological excellence, expanding their knowledge, honing their skills, and making meaningful contributions to the world of electronics.

In" this book," readers embark on a trip into the heart of electronics, exploring the essential principles of voltage, current, and resistance. This comprehensive companion serves as a roadmap for suckers and professionals likewise, furnishing a clear and terse overview of crucial generalities and real- world operations. From household gadgets to complex systems, the book illuminates the applicability of electronics fundamentals in everyday life andbeyond.

Each chapter offers precious perceptivity and practical exercises that empower readers to consolidate their understanding and apply their knowledge in meaningful ways. Whether you are a neophyte seeking to make a strong foundation or a seasoned expert looking to expand your moxie," Electronics Fundamentals" is a necessary resource for learning the secrets of electronics and unleashing new openings for invention and problem-solving.

With its accessible language, engaging exemplifications, and practicable perceptivity, this book is sure to inspire compendiums to embark on their own trip towards technological excellence. Whether you are a pupil, layman, or professional," Electronics Fundamentals" is your companion to learning the rudiments and unleashing your eventuality in the world of electronics.

www.ingramcontent.com/pod-product-compliance
Lightning Source LLC
Chambersburg PA
CBHW070311230526
45470CB00002B/826